LIt

teleph

BETWE TWO WORLDS

THE STORY OF BLACK BRITISH SCIENTIST ALAN GOFFE

Gaia Goffe
and
Judith Goffe MD

H
HANSIB

Published by Hansib Publications 2008
London and Hertfordshire

Hansib Publications Limited
P.O. Box 226, Hertford, Hertfordshire, SG14 3WY, UK

Email: info@hansib-books.com
Website: www.hansib-books.com

A catalogue record of this book is
available from the British Library

ISBN 978-1-906190-11-8

Printed and bound in the UK

For
my father Leslie
my mother Judith and
my sister Tao

CONTENTS

ACKNOWLEDGMENTS

I WOULD LIKE TO ACKNOWLEDGE THE HELP AND SUPPORT IN writing this little book of Alan Goffe's children: Tim, Jo, Carly, and Denny; and Alan's grandchildren, Bertie, Louis, TO, Alola, Jamie, Lucy-Mae, Leah, Joel, Rose, and Alana.

I would also like to acknowledge the late Elisabeth Goffe, Alan's wife and comrade, at whose cottage in Colwinston I toddled around as a toddler.

I would like to thank Alan's brother, John, for a forthright interview and much valuable information, John's son and daughter-in-law, Martin and Caroline, and their sons Daniel and Jack for all their help.

I would also like to acknowledge the help of the New York Academy of Medicine and of Alan Scadding, archivist of Epsom College, Surrey, England.

I'd like to thank Una Clarke, the first Caribbean-born woman to win a seat in New York's City Council, U.S. Congresswoman Yvette D. Clarke and her legislative director Ian Campbell, Congressman Donald M. Payne and his special assistant, Isabel Cruz.

I'd also like to take this opportunity to mention Tiffany Lim and Hillary Edelson, two of the best friends anyone could wish for.

Finally, special mention needs to be made to Alan Goffe's daughter, Jo, and to her beloved partner Alan Robertson for sifting through box after box of material to unearth amazing photos of Alan as child in Kingston-upon-Thames, of Alan in Egypt, of Elisabeth and Alan, of Ernest and Edna, and of Alan, Elisabeth and the children.

Dr Alan Powell Goffe

INTRODUCTION

THE RESEARCH SCIENTIST DR. ALAN GOFFE WAS A MAN ALL of us—whether British, Jamaican, or American—should feel proud of. He made our world a better and healthier place.

Dr Goffe's story is a fascinating one. Born in Britain in 1920 to a white English mother who was a physician and a black Jamaican father who was also a physician, Alan Goffe was one of a group of microbiologists who in the 1950s and 1960s helped develop and improve vaccines designed to fight two of the world 's most deadly infectious diseases, polio and measles.

Though Dr. Goffe's life was short, it was nonetheless filled with great discoveries and important breakthroughs. He was not concerned only with science and technology, though. He was a political person who was an early member of the Campaign for Nuclear Disarmament, the international charity Freedom From Hunger and was a member of the United Nations Association, a charitable trust. This book, *Between Two Worlds*, tells us about a man who was determined to make a difference one way or another.

I highly recommend this book to anyone who wants to understand the valuable contributions black people like Dr. Alan Goffe have made to the world of science.

Alan Goffe remains, forty years after his untimely death, a towering figure who ought to be honored and revered by people everywhere.

Yvette D Clarke United States Congresswoman
New York's 11th Congressional District
10 April 2007

───── If you have NEWS or ─────
ADVERTISEMENTS, please phone:

SEVENOAKS   54000
TONBRIDGE   4455, 4456
MAIDSTONE   58444
80 FLEET STREET   FLE 7500

Special Telephones for WANT ADS.

MAIDSTONE
55472, 55481, 55501, 58444, 53190

── Our staff is waiting to help you ──

Kent

THE C

FRIDAY AUGUST 19 1966   No. 8368

Scientist swept overboard on holiday cruise

ONE of the most outstanding virologists in Britain and a man well-known in international scientific circles was swept to his death while on a holiday cruise in his own yacht off the Isle of Wight on Saturday.

He was Dr. Alan Powell Goffe, of Sevenoaks, a vaccine specialist who made the American salk polio vaccine safe for use and was currently engaged in research on the development of a measles vaccine.

Forty-six-year-old Dr. Goffe, of 22, The Drive, Sevenoaks, was with his 13-year-old daughter and two companions when he was swept from the tiller of his 29ft. ocean racing yacht Salterello, which he had bought only a fortnight earlier.

Dr. Goffe called to his companions to loose the dinghy, but by the time the yacht was brought round Dr. Goffe had disappeared. Several yachts, including a French vessel searched, and they were then joined by a rescue helicopter and the Yarmouth, Isle of Wight, lifeboat.

But the search was unsuccessful and was called off after about five hours. Aboard the Salterello were Dr. Goffe's daughter, 13-year-old Carlena; 21-year old Mr. Simon Zachary Griffey, of 1, Linden Court, Bradbourne Park Road, Sevenoaks; and Mr. Ian Stewart-Hargreaves, 29, of Guide Square, Blackburn.

Second time

It is the second time tragedy has struck the Goffe family, as Dr. and Mrs. Goffe's 15-year-old son Hugh, a Sevenoaks School boy, died of bone cancer some years ago.

Dr. Goffe and his three companions were on a week's cruise aboard the Salterello when the accident occurred and were due to finish on the following day, Sunday. At the time of the accident Mr. Griffey and Mr. Blackburn were on the foredeck letting out a reef in the foresail.

Mr. Griffey told the KENT MESSENGER this week, "It was a moderate wind, about Force 3, with the wind dead aft. There was a very heavy swell, so much so that when we were in a trough several other yachts in the vicinity could not be seen. The boat was rolling from side to side."

Lunch

Mr. Griffey said they had been round the Needles and were inside the channel off Alum Bay, when they were working on the sail. Dr. Goffe was at the helm. They were letting out the reef when there was a sudden lurch. They looked back to see Dr. Goffe in the water, about 20 yards astern.

Dr. Goffe's daughter, Carlena, added that her father called out to them to get the dinghy overboard and let the sails down.

Mr. Griffey said that the boat was going so quickly through the water that by the time they had let the sails down and started the engine and turned back, Dr. Goffe had disappeared. The swell was so great it made it difficult to see.

Radio warning

Another yacht, Ramrod, realised there was something amiss and came alongside to ask what the trouble was. Ramrod radioed a warning back to shore and then, in company with three or four other yachts, including a French boat, they searched the area within minutes.

This was at about 1.30 p.m. on Saturday and a rescue helicopter and the Yarmouth lifeboat then joined in the search, but without success. The hunt for Dr. Goffe was called off at 5.30 p.m.

Dr. Goffe's eldest son, Mr. Timothy Goffe, aged about 20, who is a student at London Hospital, told the KENT MESSENGER that his father's holiday should have ended on the Sunday. They started by leaving Chichester at the beginning of the week and had sailed to a number of places, including Poole. His father had been interested in sailing for about a year and had gone on several courses before considering buying a yacht.

Went under

He had bought the Salterello, a 29-feet ocean racer equipped with an auxiliary engine, built at Rottmear, Sussex, in 1953, about a fortnight ago and it was his first cruise in her. There had been only two of the class built and it was "a very fast boat", Mr. Goffe added.

His father was a very strong swimmer, but at the time of the accident he was wearing a waterproof smock, waterproof trousers, and rubber boots.

Mr. Goffe felt that his father, on finding himself in the water, had probably tried to take this smock off over his head, but had gone under. Mr. Timothy Goffe had gone to the Isle of Wight on Saturday afternoon, and at first light the following day he and an uncle had gone out in a boat to search along the shore-line. They had then

DR. GOFFE

landed further along the coast and had walked back without finding any trace of his father.

Dr. Goffe had been leading a team at the Wellcome Research Laboratories in the development of a measles vaccine, with which it was believed it might be possible to eradicate measles altogether.

Colonel Hugh Mulligan, Director of the Laboratories, said this week, "This is a terrible shock. He was a most valuable scientist and one of the most outstanding virologists in the country. He had yet to reach the peak of his powers, but he was already an extremely distinguished specialist. He was a dedicated worker and a well-known figure in international scientific circles. He was well loved by his colleagues, a charming and delightful man."

Chairman

Dr. Goffe and his family came to live in Sevenoaks in 1955 and he belonged to many organisations in the town. For some years he was Chairman of the Sevenoaks branch of the Campaign for Nuclear Disarmament and last year became Chairman of Sevenoaks branch of the United Nations Association.

He belonged to Sevenoaks Film Society, was deeply interested in music and art, being a member of Sevenoaks Music Club and he also acted in productions of Sevenoaks Players and the Holmesdale Players. He was also a member of Sevenoaks Scottish Country Dancing Club, a playing member of Holmesdale Cricket Club and was keen on squash and tennis.

Both his parents were

Both his parents were doctors and after being at school at Epsom College he went to his father's old hospital, University College Hospital, London, with the idea of joining the elder Goffe's practice when qualified. He took his M.D. and B.Sc. in 1944 and then had houseman's appointments during the war at Kent and Canterbury Hospital and at the Miller Hospital, Greenwich. He then had the opportunity of taking a postgraduate course at London Hospital, where he obtained his diploma in bacteriology in 1947. He was among the last of the National Servicemen to be called up and served in the R.A.M.C. for just over a year as a pathologist, for part of the time in Egypt.

Headed team

On leaving the Army he joined the Medical Research Council at Colindale and from there went to the Wellcome Research Laboratories at Beckenham, where he was heading a team at the time of his death.

A colleague at the Laboratories commented, "Since he wished to play a part in using the newly opened opportunities of preventing disease he joined the Wellcome Laboratories in 1955, where he played a major part in the development of the Salk and Sabine vaccine against polio. His next major project concerned measles and he was the leader of the team which developed a new further attenuated strain of vaccine against it, sometimes known as the Goffe Strain, which is the only one which is a purely British development."

Tragic

"Two years ago he was appointed head of the experimental cytology unit formed to utilise Dr. Goffe's skills in critical basic research at Wellcome's. It is only eight months ago that these new laboratories were completed. His untimely death is therefore the more tragic for one will never know what his talents would have contributed to science, especially in the modern approach to the study of cytology."

Mrs. Elizabeth Goffe said on Tuesday that friends who wish to do so may send contributions to the Hugh Goffe Foundation in Sevenoaks, the trust set up in memory of their son. Dr. Goffe leaves his wife, Elizabeth and four children, two boys and two girls.

A BOY BORN TO BE
A SCIENTIST

WHEN ALAN GOFFE, A BLACK BRITISH SCIENTIST, SLIPPED AND fell into the sea while sailing off southern England in 1966, the world lost one of the most brilliant researchers of his generation. Through his work in the development of polio vaccines in the 1950s, Goffe played an important part in helping rid much of the world of the crippling polio disease.

As well as his work on the poliovirus, Alan Goffe did pioneering work on the measles virus, which had been described as "the greatest killer of children in history." So important was Goffe's work on measles, scientists named a type of the measles virus, the 'Goffe Strain', after him.

Alan Goffe was also one of the first to conduct full–scale studies of the human wart virus, which was recently discovered to be a cause of cervical cancer. In 2006, the United States Federal Drug Administration approved a vaccine that will enable young girls, before they become sexually active, to be immunised against the human wart virus. Had Alan Goffe lived he would have almost certainly played a significant role in the development of the vaccine.

Alan Goffe's death on August 13, 1966, aged only 46, was all the more tragic because he died at such a young age. His accomplishments were such that *The Times* dedicated a moving tribute to him three days after his death:

"His untimely death is therefore the more tragic," *The Times* said, "for one will never know what his outstanding talents would have contributed to science."

Alan Goffe in the 1940's.

While his death was a great loss to his family, his friends, and his colleagues, Goffe's loss would have also been felt, in some strange way, by a group of people who had never met him or probably read any of his work.

The son of a white English mother and black Jamaican father, Alan Goffe was a man between two worlds—the white world and the black world. To black people in Britain, whose lives were filled with hardship and obstacles, Goffe would have provided proof that it was possible to succeed despite discrimination. Indeed, he refuted by his achievements in the world of science, the view that people of African descent were intellectually inferior to whites.

Alan Goffe was certainly the equal, if not the superior, of most of the scientists of his generation.

He began his working life as a scientist in 1946, aged 26, as an assistant pathologist at the Bernard Baron Institute of Pathology at the London Hospital, the university hospital where he qualified as a medical doctor. He stayed at the Baron Institute for only a few months before moving in late 1946 to the Central Public Health Laboratory in Colindale, London. The next year, 1947, Alan completed a diploma in bacteriology at the London School of Hygiene and Tropical Medicine, a preparation for the extensive work he would soon begin in bacteria and viruses.

Like all young men at the time, Goffe was required to spend two years in military service. He did this between 1949 and 1951. Holding the rank of major, he served in the Royal Army Medical Corps and spent time in Egypt, where he took a special interest in the study of typhoid and other diseases. In Egypt, Goffe decided to look at how disease was spread through unsanitary handling of food. He developed a test that enabled researchers to determine whether someone responsible for handling food in a shop or restaurant or elsewhere was infected with the disease schistosomiasis and was therefore in danger of infecting others.

Major Alan Goffe in Egypt 1950

For Alan Goffe, a man of colour, the time he spent in Africa, surrounded by people who looked like him, must have been a life changing experience. His time in Egypt would probably have led the scientist to ask himself why so many black and brown people suffered and died of diseases that could be easily treated and cured with the right medicines. It was clear to Goffe, a young man with a developed social conscience, that the reason some people lived and some people died was because some were haves and some were have–nots.

Alan Goffe's time abroad must also have raised questions for him about what sort of scientist he wanted to be. Did he, for example, want to work and make discoveries that didn't necessarily end human suffering but could perhaps win him great fame, and even fortune?

It's clear he didn't want to be just another scientist. His time in Africa probably convinced him that he had a greater responsibility than did his white colleagues to help the world's suffering people as so many of them looked just like him. Goffe must have been further transformed by being in Egypt during the post World War Two period, a time of great change in the North African country. Gamal Abdel Nasser, an army officer and anti–imperialist, installed himself as president around this time. Nasser was determined to liberate Egypt from European domination and introduce major changes in the African country.

So, the times in which Alan Goffe found himself abroad as a major in the Royal Army Medical Corps, were times of great change in Africa, in Asia, in the Middle East, and later in the Caribbean, the region where his father had come from.

In 1951, his military service over with, the young scientist returned, changed by his experiences in Egypt, to his job as a bacteriologist in London. No longer a major in the British Army, he was determined now to enlist in another fight – the battle to rid the world of polio,

Major Alan Goffe in 1950

a disease that had caused suffering to millions of people since the very beginning of time.

Alan Goffe knew he couldn't change the world. But perhaps he could help make it a safer place through his work as a research scientist.

Major Alan Goffe in Egypt 1950,
second row, fourth from left

*Goffe with fellow officer in
Egypt, 1950.*

*Elisabeth and Alan
together on leave.*

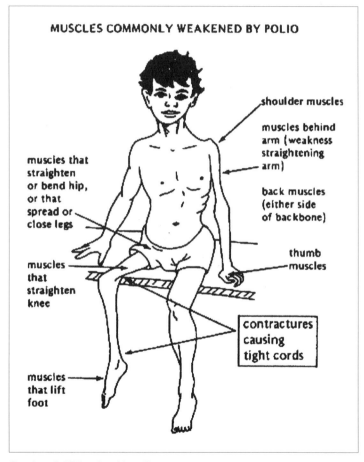

Drawing of child weakened by polio

THE THING HAS A HISTORY

POLIO DATES BACK THOUSANDS OF YEARS, TO AT LEAST AS FAR as the Pharaohs in Egypt. In the 1700s, a British physician, Michael Underwood, provided the first clinical description of the disease. Polio, Underwood said, was "a debility of the lower extremities."

Poliomyelitis is a highly contagious viral disease spread through contact with contaminated saliva or feces. The virus, somehow, finds its way into the mouth and deposits itself in the throat area. From here, it makes its way to the intestines and sheds its particles into the feces. If a person who has not been immunised or previously had the disease, somehow, makes oral contact with feces infected with poliovirus, say in a swimming pool, they will contract polio. Though it is primarily spread through fecal–oral means, poliovirus can also spread through respiratory means. For example, if someone with the virus were to cough, this would expel the particles into the air and someone could breathe it in and become infected.

Once the virus makes its way to the intestines, it can also travel to the nervous system and lead to minor symptoms such as fever or in some cases to paralysis of some limbs. In most cases, those who contract the poliovirus will not get sick, at all. Typically, only 1 or 2 people in a hundred who contract the virus will be badly affected by it. But in a population like the United States, where there are roughly 300 million people, this could result in as many as 60,000 polio cases.

Though there had probably been polio outbreaks over the years in the United States, the first known and recorded epidemic occurred in 1894. In 1908, Karl Landsteiner and Erwin Popper of

Young boy crippled by polio

the Rockefeller Institute for Medical Research in New York City first identified polio as a virus. In 1916, twenty–two years after the first reported epidemic of polio in the United States, another major outbreak occurred. Twenty seven thousand people contracted the disease and many died of it.

The event, though, that made the United States sit up and pay attention to the disease did not occur until 1921, when 39–year old Franklin Delano Roosevelt, who would one day become president, contracted polio.

Roosevelt's illness helped change the way people viewed the disease and those suffering from it. People crippled by polio had often been shunned and looked upon with disgust. This began to change, though, with the help of the Warm Springs Foundation, an organisation established with Franklin Roosevelt's help. The organisation raised millions of dollars for the treatment of polio patients and for research into the crippling disease.

It was a good thing the foundation was set up when it was. The 1930s saw a rash of new polio cases. Thankfully, in 1931, there was an important breakthrough in polio research.

Two Australian scientists, Frank Macfarlane Burnet and Annie Jean MacNamara, discovered there were two fundamental types of the poliovirus. In order to fight the virus, scientists needed to know exactly how many strains or types there were so they could then develop a method to destroy it.

In 1935, the scientific community recognised, crucially, that the way to defeat the disease was by developing a vaccine. Determined to be the first to do this, two scientists, Maurice Brodie of New York University and John Kolmer of Temple University in Philadelphia, began a fierce competition with one another to be first to come up with a vaccine. Brodie took the spinal cords of monkeys infected with the poliovirus, ground them up and added formalin, a preservative designed to kill viruses. Brodie's experiment on monkeys was a success and so he decided to test his concoction on 3,000 children.

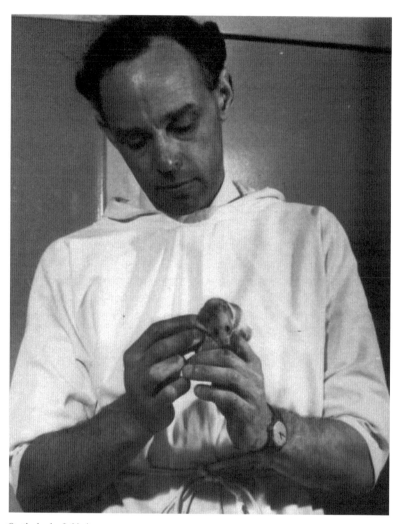

On the back of this image, someone –
probably Alan's wife Elisabeth – had
written: "I suppose in view of the very recent
fuss about animals' experiments this is not
very suitable. I know if I had to choose a
child having polio or using a monkey . . . I
would not hesitate! What do you think?"

It was a failure. Many of the children developed terrible allergic reactions. None of them developed immunity to polio by taking Brodie's vaccine.

Maurice Brodie's rival, John Kolmer, was not successful, either. Kolmer's attempts at developing a vaccine were not just fruitless, but tragic. Kolmer believed a weakened form of the actual poliovirus itself could make an effective vaccine. In other words, fighting fire with fire. Kolmer was wrong and in his haste to make his vaccine available to the public, many of the children whom it was tested on contracted the disease. Some, tragically, died as a result of Kolmer's failed experiments. After their disastrous experiments, neither Brodie nor Kolmer could find work easily and Brodie, depressed, was said to have taken his own life.

Far from advancing polio research, what Maurice Brodie and John Kolmer did was set research into the virus back many years. Few advances were made in the poliovirus field for the next decade.

World War Two proved to be both good and bad for those studying polio in the United States. Scientists were pulled off polio research by order of the US military and put to work instead on projects the government considered more important. After the war, though, when it was discovered that returning soldiers infected with polio were bringing the disease home with them, scientists were given money by the United States government to defeat the disease: Now polio had become Public Health Enemy Number One.

John Enders

THE RACE IS ON

WITH GOVERNMENT SUPPORT AND THE SUPPORT OF PRIVATE and charitable institutions available, many a breakthrough in polio research seemed very possible. In 1948, three years after the end of the Second World War, the big breakthrough finally came. John Enders, a Harvard University bacteriologist and immunologist, along with his colleagues, Thomas Weller and Frederick Robbins, came up with a revolutionary new tissue–culture technique that enabled scientists to, crucially, grow the virus in a test tube. This allowed researchers for the first time the opportunity to experiment with vast quantities of the virus. Enders' work in the cultivation of poliomyelitis virus in tissue–culture opened up a brave, new era in microbiology. Enders' innovation excited the research community and led specialists in other areas to abandon their studies to see if they could be the first to come up with an actual vaccine for polio.

On the other side of the Atlantic, in Britain, little or no polio research was taking place. For some reason even though Alan Goffe was a fully qualified bacteriologist and virologist and had demonstrated his interest in polio research, he did not make his first visit to the United States to learn Enders' tissue–culture technique until 1951. Scientists in the United Kingdom were far behind in the race to develop a polio vaccine but were running as fast as they could to try and catch up.

For some reason, only one research scientist, an American, seized on John Enders' tissue–culture technique, and made great headway with it. That scientist was Jonas Salk.

Alan Goffe in his Wellcome office.

Some people say Jonas Salk was simply in the right place at the right time. For much of his scientific career, Salk had been concerned with diseases other than polio. His main concern had been the development of a vaccine for influenza.

During World War Two, Salk was ordered by the US military to use his talents to come up with a flu vaccine. He abandoned this as soon as the war was over and he was able to pursue his own interests. Salk decided he wanted to become the first person in history to come up with an effective, safe polio vaccine. He knew if he succeeded it would make him famous.

Though scientists didn't like to admit it, they craved fame as much as professionals in other disciplines. Charles Darwin, the father of Evolutionary Biology, admitted in an interview that his interest in science "had been much aided by the ambition to be esteemed by my fellow naturalists." Max Planck, the founder of quantum physics, had expressed a similar outlook: "The scientists joy is the certainty that," Planck said, "every result found will be appreciated by specialists throughout the world."

While Jonas Salk was in the United States setting off on a path that would bring him great fame, polio research in Britain was at a virtual stand still. Alan Goffe was busy in the Central Public Health Laboratory's Virus Reference department. There he identified and logged the different types of poliovirus. It was laborious, but necessary work.

In the United States, too, scientists had begun identifying and logging different types of polioviruses. They had begun this in 1948. In 1949, scientists discovered, crucially, that there weren't simply two different types of the poliovirus, but three. The logging and identifying of types of viruses proved to be very useful, after all.

Meanwhile Jonas Salk wanted to make sure his was safe first. He knew he had time as his rivals were working principally on oral vaccines rather than the injectable one he was working on. His rival's vaccines, Salk was convinced, would not bear fruit for many, many years.

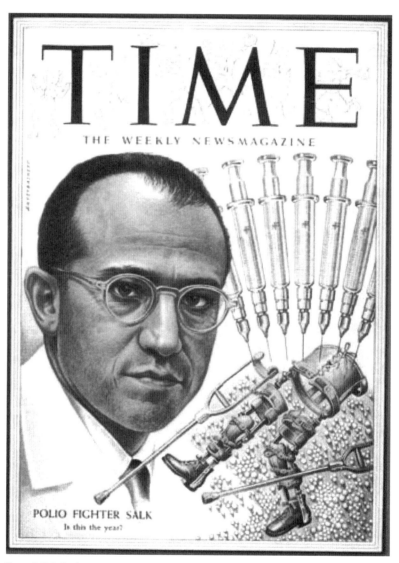

Jonas Salk is first!

Back across the Atlantic, in 1951, Alan Goffe was emerging from two long years in military service. His years in the Royal Army Medical Corps had put him behind in the polio race.

He was determined, though, to make up for lost time. He returned to the Central Public Health Laboratory, taking up the post of Senior Bacteriologist and immediately began a survey of the poliomyelitis virus in sewage by inoculating monkeys.

It was useful work. But what he really wanted to do was go to the United States to study the latest developments in polio research there. There was no time to waste, Goffe thought, if Britain was to beat America in the race to develop an effective and safe vaccine. The British government agreed and through its Medical Research Council, a body dedicated to the development of medical and related biological research, sent Goffe to find out just exactly what was going on in the United States, what Britain didn't know, and what Britain had better hurry up and find out if it was going to be competitive.

Alan shared the exciting news with his wife, Elisabeth, and with his father, Ernest. Ernest, a physician, was so thrilled his son was involved in pioneering scientific research, that he wrote a letter to his brother Alfred, in Jamaica, telling him the good news.

"Alan is being sent to the US to see what work is being done on Infantile Paralysis," Ernest wrote in August 1951. "He is sent by the government Medical Research Council with whom he is working as a bacteriologist. He will be away for a month. He is to visit the principal university laboratories in New York, Boston, Philadelphia, Baltimore, etc."

On his trip to the United States, it's likely Alan Goffe visited the Rockefeller Institute for Medical Research, a leading centre of polio research and New York University, the institution where Maurice Brodie concocted his failed vaccine in 1935. He also visited Baltimore

Alan Goffe, standing fourth from left, at
a conference in 1952.

and no doubt its famous Johns Hopkins Medical Centre. It appears he also visited Massachusetts and Pennsylvania. In Pennsylvania, there's no doubt he paid a visit to Jonas Salk's laboratory at the University of Pittsburgh. There had been talk that Salk had already come up with a vaccine and Goffe wanted to find out if this was true. His final stop was the city of Boston and the laboratory of John Enders, developer of the tissue–culture technique.

"Now the goal was truly in sight, and who got there first was largely a matter of speed," wrote the author and polio sufferer, Wilfrid Sheed.

Alan Goffe sped back to Britain immediately, and with government support, set up a tissue–culture laboratory, based on Professor John Enders' model. Goffe was appointed to a Medical Research Council committee designed to ensure scientists in Britain were aware of the very latest developments in polio research. Britain was determined to beat the United States and be the first to come up with an effective vaccine against polio. The race was on and scientists in both countries were under great pressure to produce results.

Within months of Alan Goffe's return to Britain from the United States in December 1951, news came that Jonas Salk had not only come up with a vaccine but had begun testing it on patients at the D.T. Watson Home for Crippled Children outside Pittsburgh. Salk's "killed" poliovirus vaccine appeared a success.

The tests couldn't have come at a better time. The year 1952 saw the US's biggest ever outbreak of poliomyelitis. By the end of 1952, almost 60,000 people had contracted the virus. Three thousand died as a result.

It was a bad year for the United States and a bad year for Alan Goffe, as well. In November 1952, his father, Ernest, the man who had encouraged him to become a physician and research scientist, died after suffering a stroke.

Ernest George Leopold Goffe.

Ernest and Alan had gone to an old boy's dinner at University College Hospital in London in October 1952. During the dinner, Alan thought his father seemed tipsy, or drunk. He wasn't. He was suffering a stroke. Though surrounded by some of Britain's best medical minds, none of the assorted physicians and research scientists realised what was happening to one of their own.

Somehow, Ernest, aged 85 years old and with what some called 'blood on the brain', managed to drive himself the 60 miles to his home in Ferring, in Sussex.

He took to bed and died three weeks later in a nursing home. Ernest was remembered in Jamaica's *Daily Gleaner* newspaper on December 11, 1952:

"Dr. Ernest Goffe MD (Lond), MRCS, MRCP, a member of the St Mary family, died at Ferring, Sussex, England on November 28. Born in 1867, Dr Goffe was the sixth son of the late John Beecham Goffe and Margaret Goffe of Port Maria . . . After winning the 80 pound scholarship of the year, he went to Jamaica College for one year. He then left for England to study medicine in 1889 at University College London."

Ernest and Edna Goffe with Tim, their
first grandchild, in 1946.

Ernest, left, at son Michael's wedding with Pauline; Elisabeth can be seen peeking through, third from right. Edna, Alan's mother, is third from left.

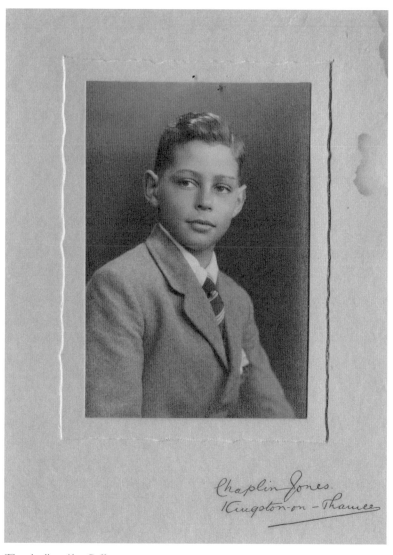

Chaplin Jones.
Kingston-on-Thames

The schoolboy, Alan Goffe

KINGSTON–UPON–THAMES, NOT KINGSTON, JAMAICA

ALAN POWELL GOFFE WAS BORN ON SATURDAY JULY 9, 1920 in Kingston–upon–Thames, a town southwest of London. He was not born in Kingston, Jamaica as one obituary wrongly reported. It's possible the writer of the obituary made this mistake because Goffe was clearly not white, and because Alan's father, a black man, was a Jamaican.

Alan Goffe's family was not typical of residents of Kingston–upon–Thames. It's likely Alan's father, Ernest, was the only immigrant and certainly the only black man living in this southern England town in the 1920s.

Ernest Goffe was a physician and surgeon and operated a thriving practice in Kingston–upon–Thames. Alan's mother, Edna, a white English woman, who was also a physician, worked with her husband in their practice.

The England in which the Goffes lived was almost lily–white. They stood out wherever they went. Though thousands of black people lived in various parts of Britain — in Cardiff in Wales, in Liverpool in northern England, in Bristol in southern England — it wasn't until the 1950s that black faces became plentiful on British streets after the British government passed legislation giving colonials from Asia, Africa, the Caribbean and other of its colonies the right to live and work in Britain.

Life for black people arriving in Britain in the 1950s was difficult. Life must have been even more difficult for Alan Goffe and his inter–racial family in lilywhite England in the 1920s.

Ernest Goffe and sons
Michael, Alan and John

In 1919, a year before Alan Goffe was born, Britain was turned upside down by race riots. Mobs of white men, recently returned from service as soldiers during the World War of 1914–1918, marched into black communities up and down Britain attacking African, Asian and Caribbean seamen whom they accused of taking 'their jobs' and consorting with 'their women.' Several were badly wounded and some were killed. There were calls to 'send the blacks back home,' though it had been the British government that had invited them to come to the 'Motherland' during the war to take up jobs left vacant by whites doing war service. Predictably, interracial relationships became a focus of white antagonism. Newspapers published ugly articles denouncing relations between black men and white women.

"Intimate association between black or coloured men and white women is a thing of horror…" said former British colonial administrator Sir Ralph Williams in 1919. American attitudes toward inter–racial relationships were well known. British attitudes were not so well known.

This was the Britain Alan Goffe was born into in 1920. Though the Goffes lived far away from the harsh conditions of Britain's inner cities, where racial antagonisms were at their strongest, they could not have failed to realise that the country in which they lived did not approve of interracial relationships.

But Alan's father, Ernest, who had come to Britain from the West Indies in the 1880s, and his mother, Edna—nearly twenty years his junior—were made of strong stuff and would not easily give in to pressure from those who disapproved of their relationship or their views of life.

Ernest George Leopold Goffe left his home in Port Maria in Jamaica in 1889 age 22 to attend medical school at University

Granny Goffe (Edna on the left with
Hugh on her lap; Elisabeth, middle;
Great Granny Powell on the right with
Jo at her knee.

Ernest and Edna's wedding
in London, 1916.

Ernest Goffe with Edna,
looking remarkably like Alan.

College Hospital in London.   His life in Jamaica had been comfortable and without racial obstacles. His father, John Beecham Goffe, a black man, had been a prosperous merchant and planter.

His grandfather, Robert Clemetson, a black man, had been a member of the island's parliament and a plantation owner. His six brothers worked in various professions. Two were attorneys, one was an engineer, and the remaining five ran the family's banana export and shipping company and at one time or another served on their parish's local government body, the Parochial Board. The Goffe family was no stranger to Britain. Three of Ernest's older brothers had been educated at boarding schools in England in the 1860s and 1870s.

So Ernest Goffe grew up, as did many black men in Jamaica and elsewhere in the Caribbean, convinced of their worth and supremely confident in their abilities. British prejudice and racism would obviously have upset and distressed Ernest, but it would not stop him.

Alan Goffe's mother, Edna Mary Powell, who was born in 1887 on the Isle of Wight in southern England, was also made of strong stuff. She came from a liberal and tolerant background that would not have encouraged her to give in to people's prejudices. Her father, Edward, was a Congregationalist minister, a Christian denomination generally thought of as progressive.

Far from opposing the union between his daughter and the black man from Jamaica who was 20 years older than his child, and once divorced, Edward Pearce Powell agreed to be the officiating minister at the wedding. The wedding, with Ernest's brother Rowland, a barrister who had lived in London since the early 1900s as best man, took place at the Kings Weigh Congregationalist Chapel in Duke Street in London in 1916.

At the time of their marriage, Ernest was employed as a resident physician at the North Eastern Fever Hospital in Tottenham,

Margaret and John Beecham Goffe,
Ernest's parents.

The successful young doctor Ernest Goffe
of the Red House, Kingston–Upon–
Thames with J.M. Barrie and friends.

London. During the First World War he treated soldiers returning wounded to Britain from battlefields in France and elsewhere. While Ernest's working conditions at the hospital in north London were quite tough, his living conditions were very plush. He lived in one of the swankiest parts of London, in Mayfair.

Edna had been living in northern England and working at the Children's Hospital in Pendlebury, Manchester. Later, she would work as a physician at the pioneering Marie Stopes Clinic, the United Kingdom's first family planning or birth control clinic. The clinic, opened in 1921 in north London, offered free birth control services to married couples for the first time.

Though from worlds apart, Ernest and Edna had much in common. They both became members of the Fabian Society, a left of centre think tank which helped inspire the creation of the British Labour Party. While the Fabians advocated reform rather than revolution, the organisation was still a big departure for Ernest Goffe from his conservative upbringing in Jamaica.

The Goffe home in Kingston–upon–Thames, known as 'The Red House', was a gathering place for intellectuals and activists of all sorts. The writer E.V. Lucas was a close family friend. J.M. Barrie, the author of the Peter Pan books, was also a family friend.

Besides his intellectual life, Ernest Goffe was a dedicated sportsman. He was a games champion while at medical school and later a member of the Rosslyn Park rugby team in London. He also acted as a referee for women's hockey on occasion. Goffe liked cricket, as well. He played on the same team and was friendly with A.A. Milne, the author of the Winnie the Pooh books.

Other acquaintances included Sylvester Williams, a fellow member of the Fabian Society. In 1906, Williams, who had been born in Trinidad, and John Archer, the son of a Barbadian man and an Irish woman, became the first black men to hold public office in Britain. Archer was elected to the Battersea Borough Council and Williams to the Marylebone Borough Council.

*A prescient image: Alan (middle) sailing
with his mother and younger brother
Michael. It was while sailing his new
yacht off the Needles—bought with
his inheritance from his mother—that
resulted in his own death in 1966.*

In 1913, Archer made history again. This time he was elected Mayor of Battersea, a district of London. It was the first time a black person had held such a position in Britain.

Raised by two individuals like Ernest and Edna, it's not surprising Alan was passionate about the sciences and social justice, as well. Alan's older brother, John, said he and his brother's background made them feel they could achieve anything. "Life was very pleasant as the son of two doctors," John said. "We did what we wanted, played sports, and everything else."

Because he had displayed an early interest in and aptitude for the sciences, Alan's father took him when he was 14, in 1935, for an interview at Epsom College, a British prep school with a long tradition of preparing the sons of physicians for a life in the sciences. Alan had a successful interview and was accepted.

But though Epsom was the ideal school for the sons of physicians who were white, it was not the right school for those who were black. The three years Alan spent at Epsom—where he was bullied and called names like 'darkie'—were the most miserable of his life. Though Alan loved his father, it was difficult for him to deal with taunts like, "your father is black, your father is a wog, your father is a nig nog."

A bright boy who had always been at the top of his class growing up, at Epsom Alan was, the college's archivist pointed out, in the bottom third of his class, barely passing subjects he knew well. The racial taunting he suffered was clearly having its effect.

Alan Goffe lived in intolerant times. During the years he was at Epsom, from 1935 to 1938, saw the rise of Mussolini's Fascists in Italy and Hitler's Nazis in Germany. In Britain, too, extremists were on the rise in the shape of Oswald Mosley's British Union of Fascists. This couldn't have made an inter–racial family like Alan's comfortable. It was a difficult time to be a racial minority in Britain.

Alan wasn't exactly black like his father, but with his dark, wavy hair and dark skin, he wasn't quite white like his mother, either.

An adolescent Alan Goffe with brothers
and friends in Kingston–on–Thames in
the 1930's.

While Alan's older brother, John, mostly escaped prejudice, Alan, John says, could not because he clearly had black ancestry.

"He had colour or racial problems because he was obviously more of coloured ancestry than perhaps I was." John Goffe said. "And also I was at a different kind of school which didn't care anything about those sorts of things."

Fed up with their son's ill–treatment, Ernest and Edna finally withdrew him from Epsom and sent him in 1938 to school in Switzerland, to the Institut Auf dem Rosenberg in St.Gallen.

Though the Nazis marched into nearby Austria while Alan was at school in Switzerland, he enjoyed his year at the institute, nonetheless. He did well enough to be accepted in 1939, the year the Second World War began, to study medicine at University College Hospital, the medical school his father had attended 50 years before.

In 1940, a year after Alan began his medical studies, news came of an enormous scientific advance. Charles Drew, an African American scientist who had been conducting experiments with the storage of blood, made a breakthrough that led to the development of blood banks. It was a great achievement, particularly for a black person living in a racially hostile society. Another African–American scientist who made great achievements in science at the same time as Drew was Percy Julian. Julian, a research chemist, was winning himself a name for his work in medicinal drugs. Alan Goffe, who was 20 years of age in 1940, couldn't have failed to notice the significance of these events. It was clear people of African descent could equal, and surpass, whites in science or in any other field given the chance.

With war raging all around him in the 1940s, Alan devoted himself to his medical studies. He interrupted them, though, in 1943 to marry Elisabeth Hedge, a daughter of the songwriter who'd written the well–known songs *Show Me The Way To Go Home* and *Try A Little Tenderness*.

Alan with Jo and Hugh in Jersey,
about 1952.

Their first child, Timothy, was born in 1945, a year after Alan completed medical school. Their second child, Josephine, was born in 1946. Their third, Hugh, was born in 1948. In 1953, Carlina arrived. Five years later, the family adopted five–year old Dennis— almost always to be known as Denny—a mixed–race boy who looked a lot like Alan. Far from trying to disguise or deny their ancestry, by adopting Denny, Alan and family seemed to be making an effort at embracing their roots.

Elisabeth and Alan with, from left, Hugh,
Tim and Jo; about 1956.

With his three brothers: from left,
Robin, John, Alan, Michael.

Back row, third from right, with his local
cricket team, Holmesdale Second Eleven,
during the 1960's.

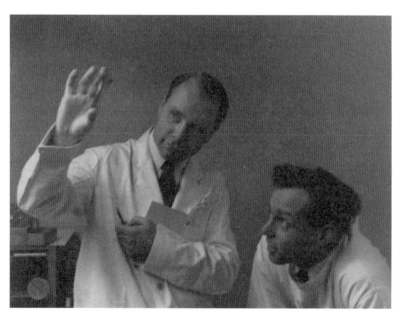

*Alan Goffe, right, with a colleague in
1952; on the back of this photograph
is a pencilled inscription in his hand:
"Toward the completion of the first final
draft! Sat., 27 ix 52".*

SALK IS FIRST ACROSS
THE FINISH LINE

NINETEEN FIFTY–TWO WAS THE BEST OF YEARS AND THE WORST of years for Alan Goffe. He published an academic paper that year in the Journal of the Royal Army Medical Corps titled *Studies on Urinary Carriage of Enteric Group Organisms*. In this paper, Goffe studied the urine of a group of food handlers, or canteen workers, to determine which one of them were carriers of schistosomiasis, a parasitic disease that affects the liver, intestines, and bladder. These parasites make their way into the feces and the urine. Once the food handlers were placed in quarantine, Goffe treated them by administering pills which removed the parasites from their bodies..

But his father, Ernest, also died that year, and Jonas Salk came up with what appeared a successful vaccine for polio that year, as well.

At first, Salk conducted trials on small numbers of children who had polio. These were successful so more trials were conducted in 1953. They were successful, as well. So Salk decided to test his vaccine on a much larger number of children. In 1953, Alan Goffe published another paper— *Poliomyelitis: Recent Developments in Laboratory Techniques*. It was in 1953, too, that perhaps the greatest scientific breakthrough ever occurred, the discovery of DNA, the very stuff of life.

In the summer of 1954 it was announced that nearly two million American children between the ages of six and nine—the so–called 'Polio Pioneers'—would be given Salk's vaccine.

Alan Goffe in his laboratory,
probably at Wellcome.

The National Foundation for Infantile Paralysis sponsored and conducted the trials, which cost the enormous sum then of nine million dollars. Nearly 30 million doses of the vaccine were given out.

There was international excitement. Newspapers speculated that a disease that had crippled millions since the beginning of time might soon be defeated.

Later in 1954, while the world waited for the results of Salk's tests, Alan Goffe and other leading virologists and bacteriologists went to Italy to attend the Third International Polio Conference.

Everyone who was anyone in polio research was there. In attendance was Albert Sabin, of the University of Cincinnati. He was convinced Salk's injectable vaccine was a bad idea. In 1952, Sabin had begun developing an oral vaccine. Alan Goffe, too, favoured this approach. Salk was, of course, there. He was, to the annoyance of his medical colleagues, the star attraction.

He told the gathering, which was packed with research scientists, physicians, public health officials, and the world's media, that two inoculations of his vaccine was all that was needed to protect a person from contracting polio. Some were supportive. Most were sceptical. They told the media they would wait until the results were announced at Christmas 1954 to offer their conclusions on Salk's vaccine.

"To wait until Christmas should not be too long," one scientist told a newspaper, "considering that we have been waiting for years before and without much hope as now."

In 1954, too, while people were waiting for the results of Salk's trials, John Enders, who had come up with the crucial tissue–culture technique, was awarded the Nobel Prize for Medicine. But no one was paying attention.

All eyes were on the United States and Salk's historic nationwide polio trials. Salk had said he had hoped to have definitive results by Christmas. The world held its breath. Alan Goffe and the other

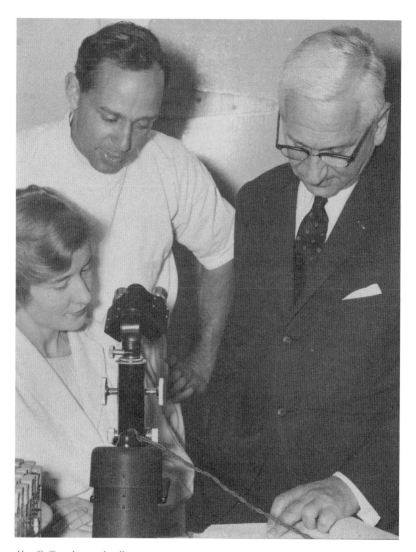

Alan Goffe and research colleague,
with Albert Sabin, at the Wellcome
laboratories, early Sixties.

scientists who had hoped to be the first to develop a safe and effective vaccine themselves were holding their breath, as well.

The world would have to hold its breath until 1955. Finally, on April 12, 1955 it was announced that Jonas Salk's vaccine was effective and safe. Within a few years, Salk's vaccine led to a drastic reduction in the number of polio cases.

In 1955, there were more than 50,000 cases of polio in the United States. By 1960, the number of polio cases had dropped to only 3,190. The next year, 1961, it dropped further to 1,312. In 1962, the number of polio cases dipped below a thousand, to just 910. Within a short time, millions of people around the world had been successfully vaccinated against polio.

It was a remarkable achievement. It made Jonas Salk a national and international hero. Though some of his rivals in the medical field remained skeptical of Salk, a man whom they said had stolen the glory from scientists who had studied the poliovirus much longer than he had, he never sought to enrich himself from his vaccine. Salk didn't patent it and so didn't earn any money directly from the millions of doses of the vaccine that were sold.

Salk's rivals, among them Alan Goffe, were left on the outside looking in. In 1955, Goffe left his government job at the Central Public Health Laboratory for a post at Britain's leading private research firm. He took up a senior position in the department of virology at the Wellcome Research Laboratories in Beckenham, England, the research and development arm of the giant Wellcome pharmaceutical company.

A colleague of Goffe's said he left his government job for one with a private company because he was disappointed by the lack of support he felt he and other scientists had been given by the British government in their effort to develop a polio vaccine. Goffe wanted to be on the cutting edge of scientific research and no one in Britain had a better record of this than did the Wellcome company.

The company was generous to its scientists, paying them well and giving them the very best facilities in which to work. Wellcome

Albert Sabin and Alan Goffe
discuss the research.

also made clear that though it wanted to make money from drugs developed by its scientists, even more important was that these drugs save lives.

At the Wellcome Research Laboratories, Goffe made major improvements to the first of Salk's poliomyelitis vaccines. While Salk's vaccine had been generally effective, all vaccines needed modifying so as to reduce possible side effects.

For seven years, between 1955 and 1962, Jonas Salk's vaccine was king. But Salk's domination of the polio research field, much to the delight of his rivals, didn't last long. His name was part of history, but by 1962 an oral vaccine made by another American scientist, Albert Sabin of Cincinnati, replaced Salk's injectable one and became the new standard for polio prevention.

By the late 1950s Alan Goffe had thrown himself into the development and improvement of the Sabin oral vaccine. He carefully studied this new vaccine and conducted trials in Switzerland and other places on it. Goffe prepared a vaccine using Sabin's method and administered it to a trial group in England. No one became ill and there were no side effects from it. It was excellent news.

While some felt Jonas Salk had rushed his vaccine into production, Albert Sabin had, by contrast, taken his time. He didn't make the headlines Salk had, but his vaccine proved much more useful to a greater number of people for a much longer period of time.

In his 2005 book, *Polio: An American Story: the crusade that mobilized the nation against the 20th century's most feared disease,* American University professor David Oshinsky said that the 1950s belonged to Jonas Salk but the 1960s belonged to Albert Sabin. In the 1960s, Sabin conducted independent vaccine trials in the Soviet Union. The USSR preferred the oral form of the polio vaccine to the injectable form. The Soviets adopted the oral polio vaccine because it was easier to administer and because it did not require the purchase of additional equipment, such as needles, syringes, swabs, and did not require trained personnel to dispense it.

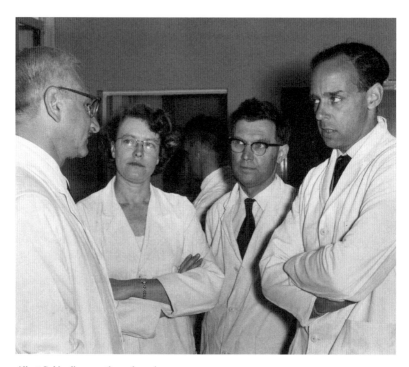

*Albert Sabin discusses the oral vaccine
with Alan Goffe and colleagues.*

In the late 1950s, both Salk and Sabin were invited to the Soviet Union by research scientists, including Mikhail Chumakov, keen to learn about the advances that had been made in polio research, both the injectable and the oral forms. Salk declined the invitation but Sabin accepted it. While on one of his visits to Europe, Albert Sabin made a special journey to Britain to meet with Alan Goffe at the Wellcome Laboratories in Beckenham, Kent. During his visit to the laboratories, Sabin talked with Goffe and his colleagues about ways of improving his oral vaccine and Goffe gave Sabin a special tour of Wellcome's state–of–the–art laboratories. The two liked one another immediately. Both men were friendly and accessible and popular with most in the scientific community. Goffe and Sabin spent many long hours at Wellcome discussing how best to improve the oral vaccine. Goffe, Sabin, and colleagues examined the poliovirus through microscopes and discussed the future of the vaccine on the laboratory steps. Happy with what he saw at the Wellcome labs and with what Alan Goffe proposed to do with his vaccine, Sabin gave his blessing to Goffe. Soon after the meeting, Goffe and his Wellcome colleagues prepared a vaccine using Sabin's method and administered it to a trial group in England . No one became ill and no one suffered side effects from it. It was excellent news.

During the same period, Sabin conducted trials on his vaccine in Cincinnati in the United States. This involved visiting schools, hospitals, and clinics each Sunday to dispense the vaccine. These outings were called 'Sabin Sundays.' John Enders and the American Medical Association gave their support to Albert Sabin by declaring that his oral polio vaccine was both safe and effective. Sabin's oral polio vaccine was finally approved by the United States Federal Drug Administration in 1963.

Though the Salk vaccine was still in use in the United States from 1963 to 1999, the Sabin oral vaccine was even more widely used. Sabin's vaccine was not without its problems. There were several cases of what virologists called vaccine–associated paralytic

With Albert Sabin on the steps of the
Wellcome laboratories; Alan Goffe is on
the left.

poliomyelitis or VAPP in the United States in the 1980s and 1990s. It was discovered that Sabin's vaccine contained traces of poliovirus strong enough to infect and cause damage to those who had taken it. As a result, in 2000 US government officials decided to return to using Salk's killed–type vaccine.

Today, however, Albert Sabin's vaccine is still the most widely used type of immunisation from polio in much of the world outside of the United States. Interestingly, despite Jonas Salk's achievement he was never elected a member of the National Academy of Sciences. It appears his colleagues – who considered him an upstart – never forgave him for becoming the first to develop and produce an effective vaccine for polio.

Alan Goffe and other protesters sitting
in the airplane at Dusseldorf airport,
having been refused entry to Germany.

A POLITICAL LIFE

ALAN GOFFE WAS A MAN BETWEEN TWO WORLDS. ON THE ONE hand, he was an employee of a large pharmaceutical company and on the other, he was committed to radical causes. While he was in his laboratory in the 1950s, working on vaccines to alleviate suffering, the world outside was deeply polarized – West versus East, Capitalist versus Communist, Black versus White.

In Africa, people were demanding independence from British rule. In America, Martin Luther King Jnr. was battling for civil rights for black people. In Britain, newly arrived immigrants found themselves under siege from racists.

The world wouldn't be changed entirely, Alan Goffe decided, by what he and others were doing in laboratories. He wanted to change the world in other ways, as well.

Raised in a progressive household and with a brother, John, a leading figure in the Communist Party of Great Britain, Alan, naturally enough, embraced progressive politics. He flirted with membership in the Communist Party, but became instead a committed member of Britain's Labour Party.

"His devotion to science was, however, only part of his whole life for he found time to follow a wide range of interests," a colleague said of Goffe. "He believed that spectacular gestures would sometimes fortify his argument."

This is true. Alan was a founding member of the Campaign for Nuclear Disarmament, CND, which was started in 1958, and

*Members of the Campaign for Nuclear
Disarmament protesting in Dusseldorf.
Alan is fourth from left.*

an affiliated organization, *The Committee of One Hundred*, which embraced the use of non–violent direct action to further the cause of nuclear disarmament.

In a demonstration of this, Alan and others participated in a sit–in on an airplane at Germany's Dusseldorf airport after the authorities blocked the protests of British and German CND marchers, by barring the British contingent entry to Germany.

When Alan realised non–violent direct action could lead to arrest and imprisonment, he returned to less confrontational protests.

"He felt he had a better chance of improving humanity," a friend said, "by helping develop vaccines to fight the most common childhood diseases and so did not want to deter that goal by being arrested."

Alan Goffe was also a member of *Freedom from Hunger*, an international charity organization that combats hunger and poverty and a member of the *United Nations Association* of the United Kingdom. Alan was dedicated to improving humanity, whether it was through politics or science.

from left: Tim, Jo, Alan and Hugh at
a medical conference dinner in Germany
in 1960.

MEASLES IS A DAMN
DULL DISEASE!

NOT ONLY DID ALAN GOFFE DO IMPORTANT WORK IN THE
development of vaccines to fight the polio virus in the 1950s, he also
did much to defeat the deadly measles virus in the 1960s, as well.

"I suppose because all of us get measles," Goffe said as he set out
to develop vaccines for the disease, "there is a tendency to regard it
as a dull and unexciting disease."

It might not have been an exciting disease for scientists to study,
but measles was highly contagious, and deadly, all the same. A
German researcher described it as the "greatest killer of children
in history."

A person who catches measles develops cold–like symptoms,
followed by an ugly rash all over their body and fever, as well.
Children are especially vulnerable to the disease. Most will recover,
but might still suffer from serious complications such as pneumonia,
bleeding disorder, seizures and coma. Even years after a person has
recovered from measles, they are not entirely safe. They can still
suffer from loss of vision, abnormal body movements, alteration in
personality, and other ailments.

Despite the terrible damage measles could cause, Alan Goffe still
felt the need to defend his decision to try and develop a vaccine to
prevent it. He did this in a paper he wrote in the 1960s:

"What is the justification for active immunisation against
measles?," he wrote. "First we must be convinced that measles is

John Enders

worth preventing: second we must be satisfied that other methods do not work and third we must show that an efficient and practical technique is available. Is measles worth preventing? Undoubtedly yes, but the urgency is greater in some parts of the world than others."

This was true. The greatest need for a vaccine was in developing countries in Africa, Asia, and the Caribbean. This, in part, explained why it was so difficult for scientists of conscience, like Alan Goffe, to convince Western governments to fund and support major campaigns against measles. The United States, Britain and other European countries rarely saw the most serious measles cases, and so didn't take it very seriously.

Still, despite governmental indifference and the indifference of researchers looking for a disease to work on that they believed would win them fame and fortune, many important scientists were committed to developing a measles vaccine.

In the United States, John Enders, the man who developed the tissue–culture technique in 1948 that allowed a vaccine for polio to be developed, found a way also to isolate the measles virus. This meant, like polio, a vaccine of some sort could be produced.

Shortly afterward, American scientists developed a weakened strain of the measles virus called the Edmonston B strain. This strain was used to make a vaccine. But the American vaccine was not free of side effects. It led some of the children who took it to contract the disease.

Because of this, Alan Goffe decided to work on improving the vaccine. First, he developed two new techniques that increased the amount of measles that could be extracted using the tissue–culture technique. He used these techniques to aid in the weakening of the Edmonston B strain. This weakening meant there was less chance for a person given a vaccine to develop the full–blown disease.

There were problems at first with Goffe's vaccine, however. In small–scale clinical trials, the vaccine Goffe was trying to weaken

Alan Goffe with Wellcome colleagues.

*Alan Goffe with government officials
and Wellcome executives.*

still proved too strong and as a result caused moderate to severe reactions in those participating in his measles trials. These reactions indicated that more work was needed before a safe, effective vaccine, without serious side effects, would be ready to be produced and made available in Britain.

While Alan Goffe and his team were busy with their rigorous testing in Britain, in the United States, in 1963, American researchers claimed they had finally come up with a safe, effective vaccine against measles.

Despite the news from America, Goffe and his team continued their experiments on the measles vaccine. In 1964, there was good news at his laboratory in Beckenham. A vaccine that Goffe's team had given to children did not cause convulsions as feared. The vaccine only caused very minor discomfort. It caused a slight fever and a slight rash. This, for the scientists, was very good news, indeed. It meant that only a little more tweaking needed to be done before the vaccine would be widely available in Britain.

What Alan discovered came to be called the 'Goffe Strain'. This strain, developed in 1961, was the only one of solely British origin.

Alan Powell Goffe

THE FINAL CHAPTER:
A GOOD LIFE

IN 1966, THE YEAR HE DIED, ALAN GOFFE HAD, IN A WAY, COME full circle in his work as a research scientist. He had spent the first years of his working life in the 1940s trying to develop vaccines to combat disease. In the mid 1960s, he found himself investigating whether scientists, in a bid to alleviate human suffering with vaccines, had, unknowingly, through experiments with monkeys, introduced other viruses, deadly ones, into vaccines.

If this was true, it meant that vaccines developed to save lives, could actually result in death in some cases. For example, the simian virus, s v 40, which caused cancerous tumours, was found to have contaminated both Salk's and Sabin's vaccines.

"He was one of the first to recognize the simian virus, s v 40, and to draw attention to the danger that it might contaminate human vaccines," wrote the leading British microbiologist, Derrick Ffarington Edwards. In fact, Goffe was one of three signers in 1961 of a letter to British medical journal, *The Lancet*, which said Salk's vaccine contained SV40, a virus that can cause cancerous tumours. Goffe's concerns were mentioned in the 2005 book by Debbie Bookchin and Jim Schumacher, *The Virus and the Vaccine: The True Story of a Cancer–causing Monkey Virus, Contaminated Polio Vaccine, and the Millions of Americans Exposed*.

Goffe's concern led to an interest in viruses that could cause cancer, such as the human wart virus. Thus, as he worked on a diagnostic test for the wart virus he speculated on its use in the early detection of cancer caused by the virus.

Alan and Elisabeth Goffe outside 22 The
Drive Sevenoaks shortly before his death.

"The development of a serological test for most virus antigen thus opens up a new avenue for the investigation of tumours in man", Goffe wrote in a paper in 1965.

Alan Goffe died before his time, of that there's no doubt. His colleagues considered him, said microbiologist Derrick Ffarington Edwards, "one of the leading virologists in Britain."

Edwards, writing in the *Journal of Pathology and Bacteriology* in April 1967, said Alan Goffe's knowledge of virology was "encyclopedic" and that Goffe was "always ready to make it freely available to his colleagues. In his approach to any problem he showed a lively imagination and great originality."

He did much in his short life. In 1965, Goffe was chosen to head a new department of experimental cytology at the Wellcome Research Laboratories. Here began the study of the structure and functions of cells in relation to immunological and other responses.

"Alan Goffe was a man of great personal charm," a colleague said, "with a wide range of interests—most, if not all, concerned with the welfare of mankind."

Thanks to him and other scientists, cases of poliomyelitis, or polio, are rarely heard of in the developed world, anymore. In fact, the Western hemisphere was declared all but free of wild polio strains in 1991. Though cases still exist in the developing world, they are, thankfully, getting much rarer. Officials at the World Health Organisation said recently, they hope the war on polio will eradicate the disease — like smallpox — in the near future.

Sevenoaks 52053

22, The Drive,
Sevenoaks,
Kent.

We are sad to tell you that Alan lost his life
in a sailing accident on 13th August, 1966.

We thank the friends who have already
written.

While the sorrow you share with us is deep,
we are also grateful and proud.

The Hugh Goffe Foundation is working well,
perhaps this is one way of remembering
Alan too.

Elisabeth, Tim, Jo, Carlina and Denny Goffe.

*Notice sent to family and friends by
Alan's surviving family.*

THE WELLCOME TRUST
(CREATED BY THE WILL OF THE LATE SIR HENRY S. WELLCOME)

52 QUEEN ANNE STREET
LONDON, W.I

TELEPHONE: WELBECK 5721-2
TELEGRAMS: WELTRUST, LONDON-WI

JSKB/PMMR.   15th August, 1966.

Dear Mrs. Goffe,

      I was shocked to read of your husband's
tragic death. I met him occasionally when I was with
the Foundation, and also more recently. We all
regarded him as one of the outstanding people working
in the research laboratories, and were quite certain that,
great though his accomplishments have been in the past,
the future held promise of even greater things.

      My fellow Trustees are all on holiday
at present, but I know they would join me in sending you
our deepest sympathy in your dreadful loss.

        Yours sincerely,

Mrs. Goffe,
22, The Drive,
Sevenoaks,
Kent.

John Boyd

Letter of condolence from
a Wellcome trustee.

Obituary Notices

A. P. GOFFE, M.B., B.S., DIPL.BACT.

Dr. A. P. Goffe, head of the department of experimental cytology at the Wellcome Research Laboratories, Beckenham, Kent, was drowned in a yachting accident off the Isle of Wight on 13 August. He was 46.

Alan Powell Goffe was born on 9 July 1920, and was educated at Epsom College, followed by a year at the Institut auf den Rosenberg at St. Gallen, Switzerland. He studied medicine at University College Hospital, London, graduating M.B., B.S. in 1944. After completing two house appointments, he went as assistant to the Bernhard Baron Institute of Pathology, London Hospital. In 1946 he accepted a post in the Public Health Laboratory Service to work at the Central Public Health Laboratories at Colindale. Almost immediately he was given the opportunity to take the course at the London School of Hygiene and Tropical Medicine for the diploma in bacteriology, and acquired this diploma in 1947. He undertook his period of National Service between 1949 and 1951, serving in the R.A.M.C. as a specialist in pathology with the rank of major. In the course of this he was stationed in Egypt.

After returning to civilian life, and to the Central Public Health Laboratory as senior bacteriologist in the Virus Reference Laboratory, he went to the U.S.A. to learn the new tissue culture techniques being developed there. The pioneer work of Dr. Enders and his colleagues in the cultivation of polio virus in tissue culture had opened up a new era in virology. Nobody realized this better than Dr. Goffe, who immediately applied their techniques to polio virus to the extent of preparing pools of virus grown in tissue culture as prototypes for vaccine production. At that time he was a member of a Medical Research Council's committee on poliomyelitis. In 1955 he joined the department of virology at the Wellcome Research Laboratories. His expressed reason for this move was to play an active part in the development of vaccines against polio and other virus diseases of man, which the newly developed techniques of tissue culture had made possible. This keen desire to use his special abilities and skills for the alleviation of human suffering was typical of him. He made important contributions to the development in Britain of both the Salk and Sabin types of polio vaccine. Next he led a team that carried out a step-wise adaptation of the Enders strain of measles virus. This led to the development of the " Beckenham strain," sometimes known as the " Goffe strain," and the only further attenuated strain of measles virus developed in Britain.

Two years ago a new department of experimental cytology was created to utilize Dr. Goffe's particular talents for original research. He set to work with energy and ... in planning this department, both in regard to its lay-out and to its long-term programme of research. It is only eight months since the newly equipped laboratories were occupied, and it can be only speculated how much cytology and science generally have lost by the untimely death of an outstanding scientist at the height of his powers. He was a Fellow of the Royal Microscopical Society, a member of the Pathological Society of Great Britain and Ireland, and of the Society for General Microbiology, and a Fellow of the Royal Society of Medicine. He contributed to a number of international conferences and was the author of numerous papers, particularly on poliomyelitis and measles. He was one of the first to recognize the monkey virus, SV 40, and to draw attention to the danger of it contaminating vaccines for man. He maintained an interest in this virus and its tumour-producing propensity, and this interest had most recently led him to studies of the human wart virus.

Alan Goffe had won himself a place among the foremost virologists in this country, and had an international reputation. At international conferences he appeared to know everybody, and foreign scientists were always welcome at his home when visiting England. He had great charm, and a particularly friendly nature with an informal manner. It is difficult for his colleagues to realize that such a vivid personality will no longer be with them. He took a great interest in general social problems, and played an active role himself from the humanitarian aspect. This part of his character was also shown in the interest he took in his staff ; on many occasions this was shown by the personal efforts he made on their behalf. Three years ago the family suffered the sad loss of a son, a schoolboy at Sevenoaks School. His family meant much to Alan and it was characteristic of him that he organized, as a memorial to his son, a charitable foundation to assist the welfare and education of children coming as sponsored students to Sevenoaks School.

Our sympathy goes out to his wife and four children. His eldest son is at present studying medicine at the London Hospital Medical School.—D.G. ff E.

C. H. GREGORY, M.A., M.D.

Dr. C. H. Gregory, formerly honorary consulting physician and honorary consulting radiologist to the Hospital of St. Cross, Rugby, died on 14 July. He was 87.

Charles Hebden Gregory was born on 14 August 1878, and received his medical education at Emmanuel College, Cambridge, and St. Bartholomew's Hospital, London. He qualified with the Conjoint diploma in 1902, and was in general practice in Whitehaven until 1903, when he graduated M.B., B.Ch. In 1904 he went to New Zealand to marry Mildred Pasley, intending to return to England after three months. Instead he acted as locum tenens in Wellington for nine months, and was then asked to go to Apiti, in the back blocks, as its first doctor. He and his wife stayed there for five years. He returned to England in 1911, proceeding M.D. the same year, and entered general practice in Aylesford, Kent, until 1914, when he joined the R.A.M.C. He was in the first gas attack at Ypres in 1915, had his wrist broken by a falling horse, and returned to England. In 1916 he was appointed deputy assistant director of medical services in the Territorials. From 1917 he was on active service in Salonika and Constantinople, until he returned home in 1919. After demobilization he joined the late Dr. Saxby in general practice in Rugby, and remained there until his retirement in 1946. His appointments included those of honorary physician and honorary radiologist to the Hospital of St. Cross, Rugby from 1927 to 1937, when he was sometime chairman and treasurer of the medical board ; he was

superintendent of Rugby Public Assistance Institution, which became an emergency hospital during the second world war, and public vaccinator. He was a founder member of the Rugby division of the British Medical Association, and chairman in 1930-2, and again in 1940-2, and was representative of Rugby and South Warwickshire division from 1934 to 1937, and again from 1941 to 1945. A member of Warwickshire panel and local medical committee from 1921, he was chairman from 1932 until his retirement in 1946. He was a member of Warwickshire insurance committee from 1936 to 1946. From its inception in 1939 he was chairman of Rugby local emergency committee.

On his retirement in 1946 Dr. Gregory settled in Godalming, and after two years became chairman of the Meath Home for epileptic women and girls, and its first president in 1966. On his diamond wedding anniversary the Meath Home nurses' cottage was opened and named after him. In 1949 he was elected to the local council as an independent, lost his seat in 1952, and was re-elected as conservative in 1953. He retired from local politics in 1962. He was mayor of Godalming in 1956-7, being elected mayor and alderman at the same meeting. In 1962 he was presented with the British Red Cross Society's badge of honour and certificate.

Charles Gregory was the senior partner of three when I joined him, and it was in this position particularly that his leadership and human qualities showed. He was highly

SELECT BIBLIOGRAPHY

ROWLAND, JOHN 1960, *The Polio Man: The Story of Jonas Salk*, Roy Publishers

PAUL, JOHN R. 1971, A History of Poliomyelitis, Yale University Press, New Haven

SEAVEY, NINA GILDEN; SMITH, JANE S; WAGNER, PAUL 1998, *A Paralyzing Fear: The Triumph over Polio in America*, TV Books, New York

PETERS, STEPHANIE TRUE 2005, *The Battle Against Polio*, Benchmark Books, New York

KEHRET, PEG 1996, *Small Steps: The Year I Got Polio*, Albert Whitman, Morton Grove, Illinois

KLUGER, JEFFREY 2004, *Splendid Solution: Jonas Salk and the Conquest of Polio*, G.P. Putnam's Sons, New York

OSHINSKY, DAVID M. 2005, *Polio: An American Story*, Oxford University Press, Oxford, New York

SHERROW, VICTORIA 2001, *Polio Epidemic: Crippling Virus Outbreak*, Enslow, Berkeley Heights, New Jersey

BOOKCHIN, DEBBIE 2004, *The Virus and the Vaccine: the true story of a cancer–causing monkey virus, contaminated polio vaccine, and the millions of Americans exposed*, St. Martin's Press, New York

ALAN POWELL GOFFE
M.B. Lond., Dip. Bact.

Dr. Alan Goffe, the virologist, was lost at sea on Aug. 13 while yachting off the Isle of Wight. He was 46.

After qualifying from University College Hospital in 1944, he served in the R.A.M.C. for two years, attaining the rank of major as a graded specialist. He then joined the staff of the Central Public Health Laboratory, Colindale, where he began a distinguished career as a virologist. In 1955 he was appointed to the rapidly expanding virus department of the Wellcome Research Laboratories, Beckenham, and he made a major contribution to its growth and development. Ten years later he became head of a newly created department of experimental cytology at Beckenham.

His special interest in research that would lead to the control of infectious diseases in man by immunising agents led him to join the Wellcome Research Laboratories. He made it clear at the time that he would be interested only in work which would have a direct bearing on the relief of human suffering. During his service with the virus department he carried out fundamental and applied research on a number of different viruses pathogenic to man, including the common cold and adenoviruses and particularly the viruses of poliomyelitis and measles. He also investigated various simian viruses occurring in tissue cultures and poultry viruses appearing as natural infections in chick embryos. Probably his contributions of greatest importance were in the development of highly effective vaccines, both of the Salk and Sabin types, against poliomyelitis and his further attenuation of Enders Edmonston B strain of measles virus which led to the development of the Beckenham (sometimes called the Goffe) strain of measles virus from which a vaccine of maximum protective value and minimum reaction was ultimately prepared and marketed. He was also closely concerned with the design and conduct of clinical and field trials with poliomyelitis and measles vaccines.

2. Read, A. E., McCarthy, C. F., Hestin, K. W., Laidlaw, J. *Br. med. J.* 1966, i, 1267.
3. Rafsky, H., Rafsky, J. C. *Am. J. Gastroent. N.Y.* 1953, 24, 87.
4. Kohler, E. M. *J. Am. med. vet. Ass.* 1964, 144, 1294.

The many publications with which his name is associated illustrate the thoroughness with which these trials and their follow-up were conducted.

A colleague at Beckenham writes:

" Goffe deservedly came to be known as one of the leading virologists in Britain, and his appointment as the English-speaking representative of *Acta Virologica* was an indication of his high international standing. His knowledge of virology was encyclopædic and he was always ready to make it freely available to his colleagues. In his approach to any problem he showed a lively imagination and great originality.

" When it was decided two years ago to create a new department of experimental cytology to study the structure and functions of cells in relation to immunological and other responses, Goffe was a natural choice as its first head. With characteristic enthusiasm he planned and equipped laboratories suitable to the needs of the highly specialised staff he had recruited to the department. It was expected that under his leadership the work of this new unit would yield information of great importance.

" Alan Goffe was a man of great personal charm with a wide range of interests—most, if not all, concerned with the welfare of mankind. He was dedicated to the relief of human suffering and he was active in the campaign for Nuclear Disarmament. His altruistic outlook was also apparent in the great interest he took in the wellbeing of his staff and in the aims and objects of the Hugh Goffe Foundation, which he and his friends established as a memorial to his son who died in 1963. He was a stimulating colleague who was liked and respected for his scientific ability and above all for his humanity."

Another friend sends this tribute:

" His gusto and his restless vivacity made him a memorable and provocative companion. Enterprise and energy he had in abundance: I wish I had a fraction of his capacity for zestful living; and I would settle for just a part of his devotion to the causes he supported. He cared ardently about the United Nations and the Campaign for Nuclear Disarmament. He believed that spectacular gestures would sometimes fortify his arguments: he sat for many hours of defiance in an aircraft impounded at Dusseldorf airport, when an exchange of British and German C.N.D. marchers was obstructed by authority. He did much for the United Nations Association. He made his friends think about issues that really mattered when they were all too ready to forget. I shall miss that stimulus."

Dr. Goffe leaves a widow and four children. A son is a medical student at the London Hospital.

*Obituary in The Lancet
20 August 1966.*

"THE HISTORY". *Global Polio Eradication Initiative.* http://www. polioeradication.org/

"FAMOUS JAMAICAN SCIENTISTS". *Pieces of the Past.* 7 July 2003. Jamaica Gleaner. http://www.jamaicagleaner.com/pages/history/ story0052.htm

ARCHER, G.T.L; GOFFE, A.P; RITCHIE, A. (1952). *Studies on Urinary Carriage of Enteric Group Organisms,* Journal of the Royal Army Medical Corps. 98, (4), 237–240.

GOFFE A. P., (1953). *Poliomyelitis: Recent Developments in Laboratory Techniques,* Proceedings of the Royal Society of Medicine. 46, (12), 1001–1004.

GOFFE, A.P., BEVERIDGE, JENNIFER, MACCALLUM, F.O., PHIPPS, P.H. (1960) *Poliomyelitis Virus in Sewage in 1951,* Bulletin of Ministry of Health and the Emergency Public Health Laboratory Service. 19, 9–18.

GOFFE, A.P. (1955) *Studies on the Distribution of Poliomyelitis Viruses in England and Wales,* Proceedings of the Royal Society of Medicine. 48 (11) 931–942.

GOFFE, A.P., PARFITT, EDITH (1955) *Quarantine Measures in Poliomyelitis,* Lancet, 268 (6875) 1172–1775.

EDWARD, D. G. FF. (1967) *Obituary,* Journal of Pathology and Bacteriology, 93 (2) 729–734.

behalf. His devotion to science was, however, only part of his whole life for he found time to follow a wide range of interests. Most outstanding was his sincere and deep-felt support of humanitarian causes, such as the Campaign for Nuclear Disarmament, Freedom from Hunger and the United Nations Association, and he was active in work for these both in his home town of Sevenoaks and elsewhere. He believed that spectacular gestures would sometimes fortify his argument. There was an occasion when he uttered a warning that his activities on behalf of the CND might put him in prison, and he was only deterred when he was persuaded that he could better serve humanity as a scientist and could contribute more by successfully producing poliomyelitis vaccine. The arts, and particularly music, were a continuing interest and relaxation. In the field of sport he was active on the squash court and cricket field. Sailing, which was to bring about his death, was the latest of his many interests.

His family life, for he was devoted to his family, provided rich background. Foreign visiting scientists always found a warm welcome to his home, where Sunday lunch provided an introduction to the English home for many others from overseas, of all races, creeds and professions. The tragic death two years previously of his son, Hugh, a schoolboy at Sevenoaks School, led, characteristically, to the opening of a memorial fund to enable children from overseas to study for a period in England.

He leaves a widow and four children, to whom we offer our sincerest sympathy. His eldest son is at present studying medicine at the London Hospital Medical School.

D. G. FF. EDWARD

BIBLIOGRAPHY

1952

G. T. L. ARCHER, A. P. GOFFE and A. RITCHIE. Studies on urinary carriage of enteric group organisms; classification of urinary carriers, and the diagnostic value of urinary antibody tests. *J. Roy. Army Med. Cps*, 1952, **98**, 40–47, 125–131, 189–197 and 237–240.

1953

Discussion on poliomyelitis: Poliomyelitis: recent developments in laboratory techniques. *Proc. Roy. Soc. Med.*, 1953, **46**, 1001–1006.

1954

G. L. LE BOUVIER, GWENNETH D. LAURENCE, EDITH M. PARFITT, MARY G. JENNENS and A. P. GOFFE. Typing of poliomyelitis viruses by complement fixation. *Lancet*, 1954, **2**, 531–532.

1955

Discussion on developments in tissue culture. Papers and discussions presented at the 3rd Int. Poliomyelitis Conf., Rome (1954), 1955, pp. 250–253.

APPENDIX

Alan Goffe's bibliography appeared at the end of the special memorial obituary published in 1967 by *The Journal of Pathology and Bacteriology* (Volume 93, No 2 pp729–734). It is reproduced, opposite, and below.

732   ALAN POWELL GOFFE

Discussion on the geographical distribution of poliomyelitis: studies on the dis-
    tribution of poliomyelitis viruses in England and Wales. *Proc. Roy. Soc. Med.*,
    1955, **48**, 937–941.
A. P. GOFFE and EDITH M. PARFITT. Quarantine measures in poliomyelitis. *Lancet*,
    1955, **1**, 1172–1175.
A. P. GOFFE, G. W. A. DICK and A. PELLISSIER. The similarity of Brazzaville
    encephalomyelitis virus and Type 1 poliomyelitis virus. *Trans. Roy. Soc.
    Trop. Med. Hyg.*, 1955, **49**, 95–96.

1956

Infections in laboratory monkeys. *Coll. Pap. Lab. Anim. Bur.*, 1956, **4**, 29.

1958

SUZANNE K. R. CLARKE, A. P. GOFFE, C. H. STUART-HARRIS and E. G. HERZOG.
    A small-scale trial of type III attenuated living poliovirus vaccine. *Brit.
    Med. J.*, 1958, **2**, 1188–1193.

1960

" In vitro " characters of virus excreted after oral vaccination with Sabin's Leon
    12alb (Type 3) vaccine. Association européenne contre la Poliomyélite,
    VIᵉ Symposium, Munich (1959), 1960, pp. 258–263.
A. P. GOFFE, JENNIFER J. BEVERIDGE, F. O. MACCALLUM and P. H. PHIPPS. Polio-
    myelitis virus in sewage in 1951. *Mon. Bull. Minist. Hlth*, 1960, **19**, 9–18.

1961

Rubeola: prospects in the prophylaxis of measles. *Proc. 6th Int. Congr. Microbiol.
    Standardization*, Wiesbaden (1960), 1961, pp. 177–190.
A. P. GOFFE, T. M. POLLOCK and F. L. SHAND. Vaccination of adults with a British
    oral poliomyelitis vaccine prepared from Sabin strains. *Brit. Med. J.*, 1961,
    2, 272–274.
A. P. GOFFE and GWENNETH D. LAURENCE. Vaccination against measles. I.
    Preparation and testing of vaccines consisting of living attenuated virus. *Brit.
    Med. J.*, 1961, **2**, 1244–1246.
P. COLLARD, R. G. HENDRICKSE, D. MONTEFIORE, P. SHERMAN, H. M. VAN DER
    WALL, D. MORLEY, A. P. GOFFE, GWENNETH D. LAURENCE and T. M. POLLOCK.
    Vaccination against measles. II. Clinical trial in Nigerian children. *Brit.
    Med. J.*, 1961, **2**, 1246–1250.
I. R. ALDOUS, B. H. KIRMAN, N. BUTLER, A. P. GOFFE, GWENNETH D. LAURENCE
    and T. M. POLLOCK. Vaccination against measles. III. Clinical trial in
    British children. *Brit. Med. J.*, 1961, **2**, 1250–1253.
A. P. GOFFE, J. HALE and P. S. GARDNER. Poliomyelitis vaccines. *Lancet*, 1961,
    1, 612.
H. D. HOLT and A. P. GOFFE. A *Corynebacterium* isolated from skin lesions in
    monkeys. *Mon. Bull. Med. Res. Counc.*, 1961, **20**, 17.

1962

M. SCHAR and A. P. GOFFE. Controlled field trial with live attenuated poliomyelitis
    virus (Sabin type) in Switzerland. Association européenne contre la Polio-
    myélite, VIIᵉ Symposium, Oxford (1961), 1962, pp. 126–136.
J. H. HALE and A. P. GOFFE. Problems of the SV40. Association européenne
    contre la Poliomyélite, VIIᵉ Symposium, Oxford (1961), 1962, pp. 151–152.

OBITUARY 733

J. M. HOSKINS, D. HOBSON, V. UDALL, TORA MADLAND, A. P. GOFFE, C. H. STUART-HARRIS and E. G. HERZOG. Small-scale trial with Sabin attenuated type 1 poliovirus vaccine in a semi-closed community. *Brit. Med. J.*, 1962, **1**, 747–753.

IRENE B. HILLARY, P. N. MEENAN, A. P. GOFFE, G. J. KNIGHT, A. D. KANAREK and T. M. POLLOCK. Antibody response in infants to the poliomyelitis component of a quadruple vaccine. *Brit. Med. J.*, 1962, **1**, 1098–1102.

Poliomyelitis—the future—prevention and control. The vaccines and their problems. *Pbl. Hlth*, 1962, 76, 221–229.

Measles vaccine. *Proc. Roy. Soc. Med.*, 1962, **55**, 846–847.

Discussion on immunization of man against measles. *Amer. J. Dis. Child.*, 1962, 103, 391–392 and 393.

1963

A. P. GOFFE, J. T. WOODALL, E. TUCKMAN, J. D. PAULETT, I. N. MANSER, L. M. FRANKLIN and P. A. L. CHAPPLE. Vaccination against measles in general practice. *Brit. Med. J.*, 1963, **1**, 26–28.

P. F. BENSON, N. R. BUTLER, J. M. COSTELLO, J. URQUHART, MOLLIE BARR, A. P. GOFFE, G. J. KNIGHT and T. M. POLLOCK. Vaccination in infancy with oral poliomyelitis and diphtheria, tetanus, pertussis vaccine. *Brit. Med. J.*, 1963, **1**, 641–643.

A. P. GOFFE and G. PLUMMER. Electron microscopy of foamy virus. *Acta virol.*, *Praha*, 1963, 7, 191.

M. SCHAR, J. LINDENMANN, H. SCHOLER, A. P. GOFFE and T. M. POLLOCK. Beurteilung der Unschädlichkeit und Wirksamkeit der Sabinschen Viren bei Massenimpfungen im Kanton Basel-Land. *Schweiz. med. Wschr.*, 1963, 93, 421–427.

1964

Active immunisation against measles. *Scient. Basis Med., A. Rev.*, 1964, pp. 151–168.

N. R BUTLER, P. F. BENSON, J. URQUHART, A. P. GOFFE, G. J. KNIGHT and T. M. POLLOCK. Further observations on vaccination in infancy with oral poliomyelitis vaccine and diphtheria, tetanus, pertussis vaccine. *Brit. Med. J.*, 1964, **2**, 418–420.

P. F. BENSON, N. R. BUTLER, A. P. GOFFE, G. J. KNIGHT, GWENNETH D. LAURENCE, C. L. MILLER and T. M. POLLOCK. Vaccination of infants with living attenuated measles vaccine (Edmonston strain) with and without gamma-globulin. *Brit. Med. J.*, 1964, **2**, 851–853.

A. P. GOFFE and ELEANOR J. BELL. A simplified intratypic serodifferentiation test: application to cases vaccinated during outbreaks. Association européenne contre la Poliomyélite, IXᵉ Symposium, Stockholm (1963), 1964, pp. 311–316.

1965

A diploid human cell strain with chronic inapparent rubella infection. *Arch. ges. Virusforsch.*, 1965, 16, 469.

J. H. HALE, A. P. GOFFE, B. E. TOMLINSON, H. R. INGHAM, M. P. WALLER and J. B. SELKON. Studies on tumours developing in hamsters following inoculation of SV40 virus and BHK21 cells. *Brit. J. Exp. Path.*, 1965, 46, 598–606.

JUNE D. ALMEIDA and A. P. GOFFE. Antibody to wart virus in human sera demonstrated by electron microscopy and precipitin tests. *Lancet*, 1965, 2, 1205–1207.

Some experiences with the Beckenham 20 vaccine. C. R. Symp. sur la Standardisation des vaccines contre la rougeole et la serologie de la rubeole, Lyon (1964, 18, 19, 20 Juin, Institut Mérieux), 1965.

734   *ALAN POWELL GOFFE*

1966

R. J. C. Harris, R. M. Dougherty, P. M. Biggs, L. N. Payne, A. P. Goffe,
   A. E. Churchill and R. Mortimer. Contaminant viruses in 2 live virus
   vaccines produced in chick cells. *J. Hyg., Camb.*, 1966, **64**, 1–7.
A. P. Goffe, June D. Almeida and F. Brown. Further information on the antibody response to wart virus. *Lancet*, 1966, **2**, 607–609.

1967

Immunization against poliomyelitis. *In* Recent advances in medical microbiology,
   ed. by A. P. Waterson, *London*, 1967, Chap. 3.

PRINTED IN GREAT BRITAIN BY OLIVER AND BOYD LTD., EDINBURGH

down at Sylvie, a sweet Golden Retriever, at my heel – 'it's something that sometimes has to be done.'

Sylvie belonged to Jack and Sadie Farrell who were occasionally overtaken by their respective work commitments and needed a hand exercising their two dogs. Sadie was a talented potter who mostly worked from home, but this week she'd been involved in an arts and crafts pre-Christmas exhibition at nearby Paddock Wood's Hop Farm.

'Sylvie isn't mine' – I reminded Kelly – 'but nonetheless I love her to bits and am happy to do the right thing by her.'

'She is a lovely dog,' Kelly agreed.

Sylvie was one of the calmest dogs I'd ever had the joy to take out. Unlike William Beagle who – along with Kelly's dog – had long disappeared amongst the woodland in search of hares and squirrels.

'ALFIE!' Kelly fog-horned, making Sylvie and me jump. 'ALFIE, WHERE ARE YOU?' She stopped and did a three-hundred-and-sixty-degree turn, scanning the undergrowth. 'Where has he gone?' she tutted. 'He's far too old and arthritic to keep up with William.'

'He'll be back,' I assured.

'Yeah, hopefully not with a dead squirrel hanging out of his mouth. I'm not good with bodies. Every time the cat dumps a mouse on the doorstep, I end up asking my neighbour to dispose of the body.'

'Doesn't Henry do it for you?' I asked in astonishment.

Kelly rolled her eyes.

'Henry doesn't "do" anything other than make me

gnash my teeth. He's drinking way too much, blaming his excessive alcohol consumption on work and stress.'

'I know,' I soothed.

'I should leave him.'

'And I should leave Derek.'

'Do you think we'll ever be brave enough to take the plunge?' she asked gloomily.

But before I could reply, William Beagle shot out of a side path. He powered towards us with Alfie in stiff pursuit.

'Oh no,' I moaned.

'Bugger,' Kelly hissed.

William had a dead hare swinging from his jaws, while Alfie had found Mr Fox's calling card and had a marvellous time rolling in it.

'Never mind our husbands,' Kelly muttered. 'Right now, we have a far more pressing problem. How am I going to clip the lead on Alfie's slime-green collar, and how are you going to persuade William to drop that furry corpse?'

Chapter Two

'*Ewww*,' wailed Kelly.

She attempted wiping her hands on a small tissue I'd found in the depths of my coat pocket.

'DON'T RUB YOUR FACE ON MY LEGS!' she shrieked at Alfie, as he proceeded to wipe himself against her denim-clad thighs. 'Oh God,' she gasped. 'Why did I ever get a dog?'

'Because they're great company,' I said, snapping the lead on William's collar.

Feeling like a murderer's dodgy accomplice, I swiftly hurled the little beagle's "gift" into the undergrowth.

Don't think about your actions, Wendy. It's not a cute bunny. It's a dead body.

'I can't believe you just did that!' Kelly blinked in horror.

'What else could I do?' I asked helplessly.

'Well, I don't know. Shouldn't we have buried it, or something?'

'Unfortunately, when I left home, I didn't think to pop a shovel in my handbag.'

'No need for sarcasm,' my bestie tutted. She dropped the soiled tissue on the woodland floor, then kicked some dead

leaves over it. 'Let's turn around and head to the public toilets. I need to properly wash my hands. I'm not looking forward to going home. There's nothing I loathe more than wrestling an enormous dog into the bathtub and trying to remove the stench of fox crap.'

'I've heard that tomato ketchup is a must when it comes to getting rid of the pong.'

'Really?' said Kelly hopefully as we headed back to the visitors' area. 'I'll give it a try. And then I'll have to wash my clothes and deep clean the bathroom before I can even *think* about bathing myself. That's ninety minutes of my life gone on a joyless task.'

'Text Steve,' I said slyly. 'See if he's able to skive off work and wash your back.'

Kelly's eyes lit up.

'I'd like nothing more than that.' She flashed me a furtive look. 'Lately, I've been fantasising about what he looks like without his clothes on. He seems to be in good shape for someone not far off fifty. Probably because of his job.'

Steve had his own construction company but wasn't averse to mucking in with the labourers. Consequently, he had broad shoulders and big biceps – from what Kelly had glimpsed when Steve had been wearing t-shirts.

The guy sounded very different to Kelly's husband. Henry had a blue whisky-nose, the paunch of a drinker, and spaghetti-skinny legs.

'Do you think it's wrong to daydream about another

man?' Kelly whispered as we passed a pair of joggers.

My answer was immediate.

'No.'

Heavens, I'd fantasised enough about men myself. Mostly actors in their heyday. Brad Pitt. Bruce Willis. George Clooney. More recently my thoughts had turned to my hunky neighbour. Ben was a whole decade younger than me with a wife who was rapidly going to seed. Lately, I'd been running a mental scenario on repeat. Ben saying, "You're locked out, Wendy? Why of *course* I'll shimmy up a ladder and climb through your bedroom window. Thank goodness you left it open." And then, when Ben reappeared downstairs, flushed and triumphant, me flashing him a smile and asking if he'd like a cuppa and a slice of cake by way of thanks. Except this was the point he'd waggle his eyebrows and say, "I have a far better idea about how you can repay me." And naturally I'd oblige. Energetically. Joyfully. Enthusiastically. Him sweeping me into his arms. Me swooning prettily. Him lowering his mouth to mine. Me closing my eyes as the tip of his tongue–

'Why are you panting?' asked Kelly.

My thoughts scattered like confetti.

'Er, because we're walking faster than usual. It's making me a bit, you know, out of breath.'

Kelly narrowed her eyes.

'Hm.'

'Anyway,' I chirped, keen to avoid her scrutiny. 'I expect sooner or later you and Steve will take things to the

next level. I mean, you can't keep lurking in Lidl or skulking in Sainsbury's.'

My friend was instantly distracted, as I had known she would be.

'You're right,' she sighed. 'But it's one thing to keep flirty company and have a stolen kiss. It's quite another to do the deed. And anyway, where would we go? I daren't risk inviting him back to my place. I have two savvy teenagers who have a habit of turning up when least expected.' Kelly momentarily stared up at the sky as she visualised a scenario. 'Imagine. Ten o'clock in the morning. Henry safely ensconced behind his desk at the office. The boys at their respective universities and conveniently out of the way. And then, just when I'm stripped down to my undies and my lover is admiring the way my boobs haven't quite yet reached my navel, my sons bursting into the bedroom and reminding me they're home for study days and' – she adopted a shocked accent – '"Good GOD, Mother! Why are you breastfeeding a strange man?"' She tore her eyes away from a flurry of cumulus clouds. 'No, it would never work. Anyway, I have nosy neighbours.'

'Everybody in Little Waterlow has nosy neighbours,' I pointed out.

'True,' she agreed.

'What about going to Steve's place?' I suggested.

'Definitely not.' Kelly gave a mock shudder. 'Caroline is a full-time stay-at-home wife who is kept *extremely* busy supervising the cleaning lady, overseeing the gardener, and

standing over the ironing lady who apparently even presses the family's pants.' She looked at me incredulously. 'I mean, *who* irons pants?'

'I have no idea,' I said, flushing slightly as I recalled another fantasy. Me standing over the ironing board. Blasting steam everywhere as I ironed black cotton briefs belonging to Yours Truly. The radio playing. Tom Jones belting out *Baby You Can Keep Your Hat On.* Me changing the lyrics and singing, "Wendy you can take your pants off," just as Ben magically appeared in front of the ironing board and hooked his fingers through the belt-hoops of my jeans... zipper going down... top coming off... until I was standing before him in nothing but my one good lacy bra and a pair of pristine briefs – not a crease to be seen – which miraculously also reflected my face on this occasion. Yes, I thought fervently, recalling my ironed pants. Best to be on the safe side. These days one needed all the help one could get to look good at forty-eight. If that meant ironing one's pants, then so be it.

'Wends? Are you listening to me?'

'Of course I'm listening to you,' I said, snapping to. 'Caroline sounds like a complete nightmare.'

'She is.' Kelly nodded fervently. 'I reckon if she caught me and Steve at it, instead of bawling him out for adultery, she'd more than likely march over and yell, "Not like THAT, you stupid man!" Apparently, she is a complete control freak in all areas, so I can't imagine her being any different in the bedroom.'

'Are they still sleeping together?' I asked carelessly. Kelly's face crumpled, and I realised I'd fluffed up by asking such a stupid question. 'Sorry, that was insensitive of me.'

'It's fine.' Kelly did a few rapid blinks and composed herself. 'Steve says not. He says they haven't done anything for yonks.'

'And do you believe him?'

She shrugged.

'I want to. He's asked me too. You know… whether Henry and I are still active under the duvet.'

'And are you?'

She looked anguished for a moment.

'Well, generally, no.'

'What do you mean *generally* no?'

'I mean, definitely no. Not for ages. Until two nights ago.'

'What happened two nights ago?'

'Something crazy. We're talking totally whacko. I'd eaten a load of blue cheese before bedtime and… did you know that cheese consumed in the evening can make you vividly dream?'

'I didn't.'

'Well, it does. And it did. And…' Kelly trailed off awkwardly.

'Ah.' Realisation dawned. 'You'd eaten blue cheese and found yourself having *blue* dreams.'

'Exactly!' she said, grateful for me cottoning on. 'So, there I was. In bed. Fast asleep and having the most

incredibly heightened dream starring the delectable Steve and moi. Unsurprisingly, cheese featured. Steve wanted to play silly games. He'd dressed up as a pony with a mask on and I had to guess which cheese he was.'

'Mascarpone?'

'Very good, Wends! Well, the dream progressed, and we were having this bizarre conversation about what sort of music a cheese would listen to-'

'R & Brie?'

She nodded.

'And then Steve was whispering in my ear – which seemed incredibly erotic at the time – and he was saying that the Big Cheese had turned up and needed handling-'

'Caerphilly?'

Kelly gave me a sharp look.

'Were you having the same dream as me last night?'

'Absolutely not,' I said hastily.

'Anyway. Steve began nuzzling my neck and I let out a blissful sigh as he cooed, "I'm right brie-hind yoooo!" But… but…'

'But what?' I prompted.

'But it wasn't Steve behind me. It was Henry. In real life. My eyes snapped open in the dark just as Henry – boozy breath billowing – took advantage of his conjugal rights with a compliant wife who'd apparently woken him up moaning, "Do it! Do it! I Camembert it any longer." Henry has since stocked up the fridge with three vast blue cheeses and spent yesterday evening encouraging me to have a bedtime snack.'

'Oh dear,' I said, trying not to giggle.

'Do you think that counts as cheating?' she asked fretfully.

'Hardly. After all, Henry is your husband.'

'Not with Henry,' Kelly tutted, rolling her eyes. 'I meant with Steve.'

'No,' I said, telling her what she wanted to hear. 'You're not in a proper relationship with Steve, so how can you have cheated on him?'

'Yes,' she said, sighing with relief. 'You're right. Phew.'

'We're here.' I pointed to the Bluebell Café and the public loos alongside. 'Let's clean up and then have a quick cuppa before heading home.'

'I don't think I have the nerve to go inside the café reeking of *Eau de Monsieur Renard.*'

'I'll do it,' I said, stooping to the task of looping three sets of dogs leads around a nearby post-and-rail fence. 'Let me quickly wash my hands first. Do you want a Cheddar toastie too?'

'No thanks,' said Kelly. She flashed me a grin. 'Right now, I'm all cheesed out.'

Chapter Three

Ten minutes later, we were sitting outside the café, bums on a bench seat, drinks resting on one of the country park's trestle tables and doing our best to ignore the stiff November breeze.

We huddled into our coats whilst admiring the surrounding woodland which was a stunning backdrop of red, orange and gold. The grassy area around our table was heavily coated in leaf fall, indicating winter was just around the corner.

The three dogs sat together, looking on. Alfie and Sylvie were behaving impeccably, but William was having none of it. He kept up a steady stream of barking, baying his displeasure at being excluded from sampling a slice of carrot cake.

'*Arrrooooooo!*' he wailed repeatedly.

We'd already received some pained looks from other walkers who'd paused to enjoy an al fresco coffee and snack.

'Ignore,' I said to a rattled Kelly.

'Who? William or that cross looking elderly couple sitting over there?'

'Both,' I said firmly.

'I know those two,' she hissed, leaning in.

'Me too,' I whispered back.

'Fred and Mabel Plaistow. Little Waterlow's oldest residents who think they're entitled just because they're days away from receiving the Queen's telegram.'

'You mean the King's telegram,' I pointed out. 'Anyway, I don't think they're *that* ancient.'

'Mabel is giving us absolute daggers,' Kelly muttered. She turned and gave them her own ferocious scowl. 'There,' she said, smirking with satisfaction. 'That'll teach them to glare. Now then' – she took a sip of her drink, a regrouping gesture – 'this morning all I did was moan about Henry. Now it's your turn to moan about Derek. Tell me what's going on.'

I took a slurp of my own drink before replying.

'Nothing,' I sighed. 'It's the same old same old.'

'He's still picking his nose?'

'Yes.'

'And hogging the remote control?'

'Yes. But it's not his habits that rattle the chains of our marriage, Kelly.'

'Ah. In other words, he's still a self-opinionated, arrogant prat who treats you like the dog poo you pick up.'

'Yes,' I sighed again.

'Why don't you leave him?'

'Why don't you leave Henry?' I countered.

'I will when the boys have graduated from uni. I can't risk jeopardising their exam results. Do you *really* think you'll ever leave Derek?'

'One day,' I said, but my words lacked conviction.

'When exactly is *one day*?'

'Not sure. Anyway, an escape needs proper planning. After all, I have responsibilities.'

My friend rolled her eyes.

'You mean being a doormat.'

'That's a bit harsh,' I winced, trying to not look hurt. 'Ruby might not be at uni like your boys, but she's still living under our roof. Having unexpectedly made us grandparents, I can hardly turf her out along with my granddaughter, can I?' I felt upset at the very thought. 'I know your lads gave you the run-around for a while, but at least they got through their early teenage years without falling in with the wrong crowd and ending up pregnant at fifteen.'

'Well, being male, they could hardly get pregnant,' Kelly pointed out.

'No, but they could easily have got someone else's daughter up the duff, and how would you have felt then?'

'Awful,' Kelly admitted. 'I'm sorry. Now it's my turn to have dropped a clanger.'

The subject of Ruby was hardly a fresh one between me and my bestie. A little after her fifteenth birthday, my daughter had announced that her pal Simon was more of a mate than Derek and I had realised. Discovering that a missing ten-pound note from my purse had paid for a pregnancy test – one that showed two blue lines – had given me an instant out-of-body experience.

At the time my mind had splintered in two. I'd watched, from somewhere around ceiling level, the scenario that had gone on to unfold in the kitchen.

There was Ruby in her school uniform. And Simon, tall and gangly. Derek, mercifully, had not yet arrived home from work, so was not around to punch Simon's lights out.

On that lifechanging morning, Ruby had left for school with her hemline too short, eyes defiantly outlined in kohl, and her newly dyed hair an interesting shade of purple. Her attitude for the last twelve months had seesawed between truculent and sneering.

Back then, she'd looked down her nose at me. Her expression could shift from pity to disgust faster than a Tesla's acceleration. I was aware that raging hormones were behind the transformation in my previously sweet girl. Almost overnight she'd morphed into a spotty, bad-tempered adolescent with unexpected tantrums fuelled by premenstrual tension. Suddenly everyone was an idiot. I was an idiot. Derek was an idiot. Even the cat next door was an idiot.

On this particular day, she'd returned home from school with Simon. Ruby had revealed an emotion never witnessed before. Fear. She'd tried to disguise it by carelessly tossing the plastic tester across the kitchen table but had then slumped down on a chair. I'd not needed any further explanation.

'I'm not having an abortion,' she'd shrieked into the shocked silence. 'Simon and I have already discussed it. We

want to be together. If we have a boy, he'll be called Sunshine Lord, and if it's a girl, she will be named Moonbeam Fairy.'

I'd been unable to reply on account of my essence bobbing somewhere around the kitchen's lightshade. I'd looked on – like a member of the audience at a play – as the drama played out.

Ruby's boyfriend had decided not to sit down at the table – possibly in case Derek came home early and Simon needed to make a quick getaway. The kitchen was too small to pace around, so instead he'd had a complete outbreak of the fidgets, shifting his weight repeatedly from one foot to the other.

'SAY SOMETHING!' Ruby had roared, making Simon visibly jump.

Her outburst had catapulted my mind back into my body which had been leaning against the worktop. The edge had dug painfully into the small of my back.

'Erm,' I'd croaked, before hastily clearing my throat. 'Nice names.'

I hadn't dared say otherwise. Who knew what wrath my disagreement might have caused? Yes, I admit it. I'd been a little scared of my teenager. Keeping the peace had been paramount. Avoiding rows, essential. Deflecting temper outbursts had been all. And trying to keep Derek and Ruby at arm's length from each other had been crucial because, if father and daughter went head–to–head, I'd inadvertently end up as whipping boy for the pair of them. Pathetic? Yes,

maybe. But unless you've shared your home with a human box of fireworks, you won't know what it's like being permanently on edge. Waiting for one of them to light the emotional fuse – BOOM! – leaving you trembling at the vileness. The hateful words. The breathtaking ugliness of it all.

Want to know what happens next?
You can download the book from Amazon

Also by Debbie Viggiano

Wendy's Winter Gift

Sophie's Summer Kiss

Sadie's Spring Surprise

Annie's Autumn Escape

Daisy's Dilemma

The Watchful Neighbour (debut psychological thriller)

The Man You Meet in Heaven

Cappuccino and Chick-Chat (memoir)

Willow's Wedding Vows

Lucy's Last Straw

What Holly's Husband Did

Stockings and Cellulite

Lipstick and Lies

Flings and Arrows

The Perfect Marriage

Secrets

The Corner Shop of Whispers

The Woman Who Knew Everything

Mixed Emotions (short stories)

The Ex Factor (a family drama)

Lily's Pink Cloud ~ a child's fairytale

100 ~ the Author's experience of Chronic Myeloid Leukaemia